AF228466

VANILLA

by Golriz Golkar

Cody Koala

An Imprint of Pop!
popbooksonline.com

abdobooks.com
Published by Pop!, a division of ABDO, PO Box 398166, Minneapolis, Minnesota 55439. Copyright © 2022 by Abdo Consulting Group, Inc. International copyrights reserved in all countries. No part of this book may be reproduced in any form without written permission from the publisher. Cody Koala™ is a trademark and logo of Pop.

Printed in the United States of America, North Mankato, Minnesota.

052021
092021

THIS BOOK CONTAINS RECYCLED MATERIALS

Cover Photos: Shutterstock Images, foreground, background
Interior Photos: Shutterstock Images, 1 (foreground), 1 (background), 16; iStockphoto, 5, 7 (bottom left), 7 (bottom right), 8, 11, 14, 17, 19 (top), 19 (bottom left), 19 (bottom right); Red Line Editorial, 7 (top); Owen Franken/Corbis Documentary/Getty Images, 13; Leonid Iastremskyi/Pixel-shot/Alamy, 20

Editor: Aubrey Zalewski
Series Designers: Laura Graphenteen and Colleen McLaren

Library of Congress Control Number: 2020948887
Publisher's Cataloging-in-Publication Data
Names: Golkar, Golriz, author.
Title: Vanilla / by Golriz Golkar
Description: Minneapolis, Minnesota : Pop!, 2022 | Series: How foods grow | Includes online resources and index.
Identifiers: ISBN 9781532169830 (lib. bdg.) | ISBN 9781098240769 (ebook)
Subjects: LCSH: Vanilla--Juvenile literature. | Cooking (Vanilla)--Juvenile literature. | Vanilla industry--Juvenile literature. | Agriculture--Juvenile literature. | Food crops--Juvenile literature.
Classification: DDC 631.5--dc23

Hello! My name is

Cody Koala

Pop open this book and you'll find QR codes like this one, loaded with information, so you can learn even more!

Scan this code* and others like it while you read, or visit the website below to make this book pop.

popbooksonline.com/vanilla

*Scanning QR codes requires a web-enabled smart device with a QR code reader app and a camera.

Table of Contents

Chapter 1
What Is Vanilla? 4

Chapter 2
Growing a Vanilla Plant . . .6

Chapter 3
Preparing Vanilla Pods . . 12

Chapter 4
That Tastes Like Vanilla! . . 18

Making Connections 22
Glossary. 23
Index 24
Online Resources 24

What Is Vanilla?

Vanilla is a spice. People use it to flavor many foods. Vanilla comes from a **tropical** plant. A vanilla plant has flowers on long vines. It needs lots of care to grow.

Watch a video here!

Chapter 2

Growing a Vanilla Plant

The vanilla plant is **native** to Mexico. Today it mostly grows in Tahiti and Madagascar. Farmers grow vanilla on **plantations**.

Where Vanilla Grows

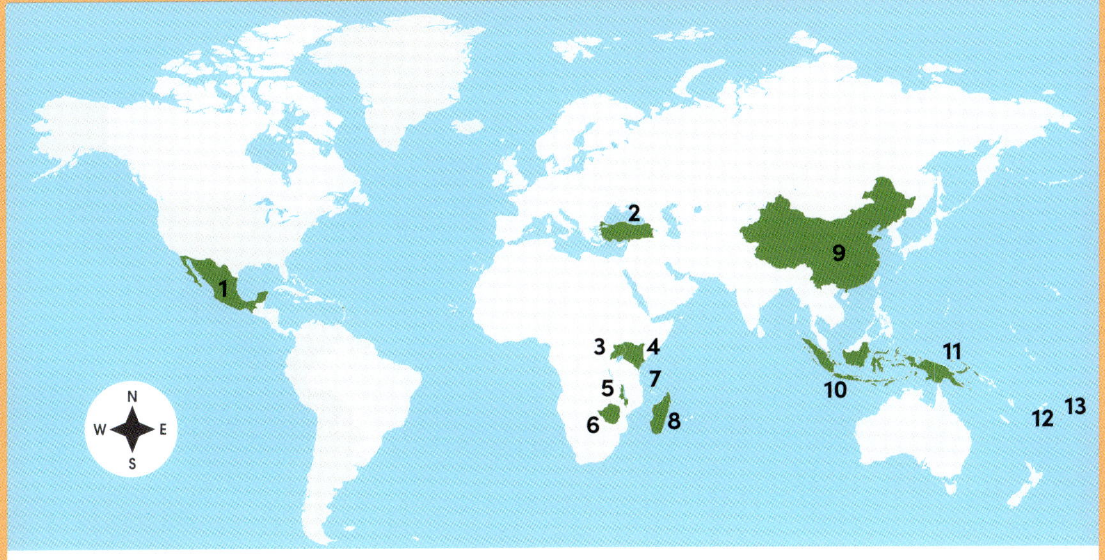

1. Mexico
2. Turkey
3. Uganda
4. Kenya
5. Malawi
6. Zimbabwe
7. Comoros
8. Madagascar
9. China
10. Indonesia
11. Papua New Guinea
12. Tonga
13. French Polynesia

Learn more here!

Vanilla flowers bloom once a year. They stay open for one day. Vanilla flowers cannot **pollinate** on their own. Farmers do it by hand with small, pointed sticks. Only pollinated flowers can grow vanilla pods.

Once pollinated, a flower's thick, green base swells. It grows a pod full of tiny seeds. Farmers pick the ripe pods after approximately nine months.

flower
pods
vine

Chapter 3

Preparing Vanilla Pods

The picked pods are cured.

Curing dries out the pods.

It strengthens their flavor.

Curing takes many weeks.

First, the pods are placed

briefly in hot water.

Complete an activity here!

DOM 358

Next, the pods are kept warm and moist in boxes at night. This is called sweating. During the day, they dry in the sun. The pods go between sweating and drying for up to two months. They turn dark brown and release an **aroma**.

The pods' final drying
stage is in an airy room.
Then they are ready to sell.

The Totonac and Aztec peoples of Mexico
were the first to grow and use vanilla.

Vanilla may be sold as a
pod, a powder, or a liquid.
The liquid is called extract.

Chapter 4

That Tastes Like Vanilla!

Vanilla is a common flavor for ice cream, cakes, and cookies. It is an important **ingredient** in chocolate and candy.

Learn more here!

Vanilla
EXTRACT

People use vanilla when baking at home. Recipes often need vanilla extract. It adds flavor to sweet treats.

Making Connections

Text-to-Self

Have you eaten something with vanilla as an ingredient? Did you like the flavor?

Text-to-Text

Can you think of another food you have read about? How is it like the vanilla plant? How is it different?

Text-to-World

Vanilla grows in tropical places. What are some tropical places that you know of?

Glossary

aroma – a strong and usually pleasing smell.

ingredient – one substance used in a mixture.

native – naturally living in a certain area.

plantation – a large area of land where one type of plant is grown.

pollinate – to move pollen from the male part of a flower to the female part.

tropical – having to do with a place where the weather is usually warm and wet.

Index

drying, 15–16

extract, 17, 21

flowers, 4, 9–10, 11

plantations, 6

pods, 9–10, 11, 12, 15–17

pollination, 9–10

sweating, 15

vines, 4, 11

Online Resources

popbooksonline.com

Thanks for reading this Cody Koala book!

Scan this code* and others like it in this book, or visit the website below to make this book pop!

popbooksonline.com/vanilla

*Scanning QR codes requires a web-enabled smart device with a QR code reader app and a camera.